Arnold Lawson, a retired teacher, was the head of biology and head of the science faculty in a large comprehensive school in Sheffield. From an early age and living on the edge of the English Lake District with its outstanding landscapes and its diversity of flora and fauna, he has always had an interest in natural history. This interest has taken him from the forests of New Zealand to the Amazonian jungle of Brazil. His other interests include gardening, fell walking and landscape photography.

For Madeline, Kathryn and Hazel

Arnold Lawson

HORRIBLE BIOLOGY

AUSTIN MACAULEY PUBLISHERS™

LONDON * CAMBRIDGE * NEW YORK * SHARJAH

A CIP catalogue record for this title is available from the British Library.

ISBN 9781398495005 (Paperback)
ISBN 9781398495012 (ePub e-book)

www.austinmacauley.com

First Published 2024
Austin Macauley Publishers Ltd®
1 Canada Square
Canary Wharf
London
E14 5AA

The author wishes to acknowledge with greatest thanks, the encouragement, kindness and assistance the production team of Austin Macauley Publisher. I also wish to thank the help given by wife, Madeline, in encouraging me in times of Covid-19 to continue to research and write the book. Also, to Hazel, for sorting out computer problems and Kathryn, for reading and suggesting changes to the manuscript and topics.

Table of Content

Introduction

Do you know how many different kinds (or species) of living things (or organisms) there are on earth? You don't know? The author doesn't know and neither does anyone else! There about 1.2 million ANIMAL species, but biologists believe there could be as many as 8 million different species. New animal species are found every year. In 2018, biologists discovered 120 new species of wasp, 34 sea slugs, 19 fish, 7 spiders, 3 sharks, 1 frog and 1 snake. In 2021, over 550 new species were discovered, including 52 new species of wasp, 7 crabs, 6 flies, 13 moths, 91 beetles and 10 new species of amphibians and reptiles. A new species of carnivorous plant was discovered and named the killer tobacco plant, and another animal was the screaming tree frog. However, the UN (United Nations) have suggested that between 150–235 species become extinct every day due to climate change and the destruction of the environment.

So far, 10,500 species of bird have been identified, 33,600 species of fish, 18,500 species of butterfly, 2,500 fleas and 2,500 species of earthworm. There are two main reasons why many species remain undiscovered. Firstly, many live in very inhospitable areas, for example the bottom of deep oceans or unexplored areas of jungle, and secondly, many are extremely small and microscopic in size.

Another group of organisms are PLANTS which include about 390,000 different species. In 2016, 2,000 new species of plant were discovered.

Another group is called FUNGI, and this includes mushrooms, toadstools, and yeasts and about 75,000 different species have been identified.

Another group of organisms is called PROTISTA, most of which are very small and over 200,000 species have been identified.

BACTERIA make up another group and biologists think there are between 10 million and 1 billion different species. In a study carried out by the Marine Biological Laboratory in America, 20,000 different species of bacteria were found in one litre of seawater!

So altogether, there are five large groups of organisms – ANIMALS, PLANTS, FUNGI, PROTISTA and BACTERIA, and each species must belong to one of these groups.

Obviously, most of these species are harmless. Some are annoying, but there are species that are dangerous, some very dangerous. Most people think, for example, of lions, tigers, sharks, and cobra snakes as being dangerous. However, most people would not think of mosquitoes as being dangerous and yet they are responsible for more deaths in the world than any other organism. Strangely, the mosquito does not kill – it is the female mosquito that feeds on blood and if that person is infected with a microscopic malarial causing organism, then when the mosquito feeds on another person, this small organism is injected into another person and so the disease is spread. These mosquito-borne diseases kill almost a million people a year.

In this book, you can read about species of animals, plants, fungi, protista and bacteria that can cause serious harm and even death. Some of the effects of these organisms on people are horrendous and billions of dollars are spent each year by the World health Organisation (WHO) and other organisations in trying to eradicate these organisms and improve the health of people throughout the world.

However, not all diseases are caused by animals. Many people suffer from the effects of poisons found in a range of organisms, and a few plants capture and eat animals. James Lind, Stephen Hales, Samuel Pepys, Alexis St. Martin and Douglas Mawson contributed to our understanding of biology and their stories and experiences are sometimes horrible but fascinating.

1. Demodex

When you are asleep, Demodex comes out of its hiding place and crawls over your face. Demodex are very small and are about 0.3 to 0.4 mm in length. Photographs of these very small spiders are shown on the opposite page. They live in the roots of your facial hairs and are also found in the roots of nasal and eyelash hairs. They feed on dead skin cells and on oils produced by the skin. They are very common, and most people do not know that they even have them, although they are more common in older people. About a third of children, half of adults and two thirds of older people have Demodex.

Demodex is a very small spider, known as a mite and has eight legs, and because it lives on the roots of the eyelash, it is sometimes called the eyelash mite. There are both male and female Demodex, and after mating, the female lays its eggs on the root of the hair. They do not like light and only travel on the surface of the skin when it is dark where they can crawl across the skin at about 16 mm per hour. Because they are so small, you do not feel them crawling about and they will certainly not wake you from your sleep. In fact, they are almost harmless, and it is only if you have a very large number of them, are they likely to cause you some harm. It is thought that the skin disease called acne may be connected to increased numbers of the mite.

They were first discovered in 1841, but the discovery was ignored, and it was a year later in 1842 when Demodex was scientifically described by a German called Gustav Simon, who was a surgeon in the army and worked in army hospitals. He was the first person to find out that somebody could survive with only one healthy kidney. He also researched the skin disease called acne and when he placed some of the pus from the acne sore under the microscope, he found minute white "worms" which were eventually found out to be the mites we now call the eyelash mite or Demodex.

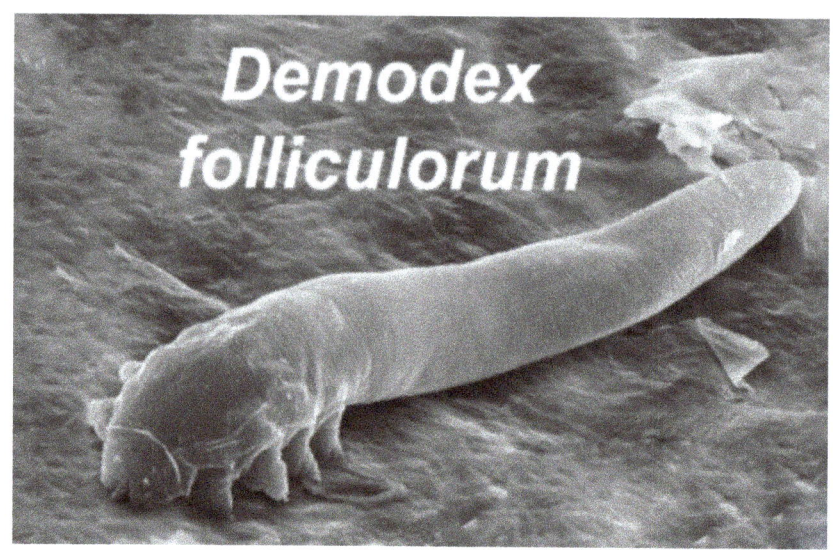

1 A Demodex as seen under the microscope. On its left-hand side four legs can be clearly seen. With four legs on the right-hand side, a total of eight legs classifies it as a spider!

2. A Demodex has been computer coloured to make it a little clearer to see.

3. These Demodex spiders have been stained to make them clear. The vertical Demodex are burrowing down into the microscopic holes where the eyelashes have been growing.

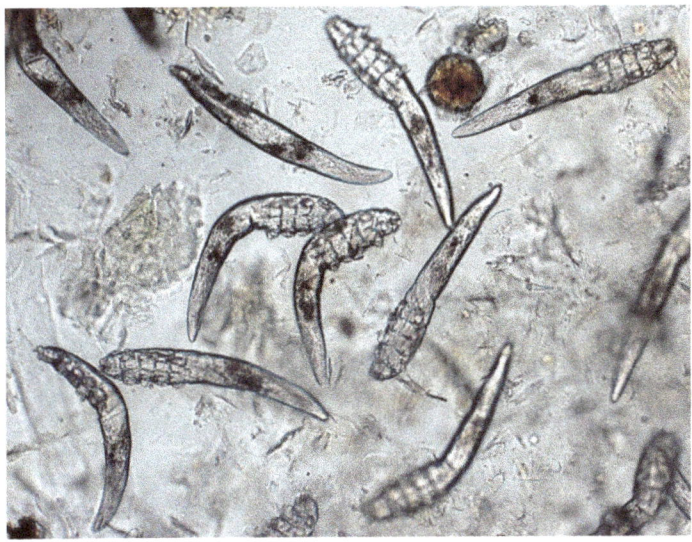

4. The eyelashes of a person with a serious infection of Demodex was wiped and the smear was placed under the microscope and photographed

2. Captain James Lind and Scurvy

How would you have liked to be a sailor on the sailing ship HMS Salisbury which is shown on the opposite page? Perhaps it would have been quite exciting, but you would have been on board for three months, six months or even a year! It would have been a very hard life – cold, wet, climbing the rigging and with very poor food. This is what it would have been like centuries ago on a sailing ship. One of the dreaded ailments that sailors feared the most was a disease known as scurvy. It usually occurred when sailors had been living on a sailing vessel for about three months. The sailors would have been very ill. They would have suffered from very soft bleeding gums, bleeding under the skin, large red, black bruises to the skin and raised lumps around the hairs on the legs. Probably worst of all was the smell caused by the rotting of their own flesh. Eventually, the sailor would die.

James Lind was born in Edinburgh in 1716 and in 1731 started his medical studies in the same city. In 1739, he entered the British Navy as an apprentice surgeon and in 1847, he was appointed as the doctor and surgeon on HMS Salisbury. While he was sailing on HMS Salisbury, 90% of the crew died from the dreaded scurvy. Because so many sailors were dying on the ship, he carried out one of the most important experiments in medical history.

In his experiment, he used 12 sailors who were suffering from the disease and divided them into six groups of two. Each group was given a different drink with their food and this drink was given to them for fourteen days. One group was given one litre of cider a day, the second group was given 25 ml of dilute sulphuric acid, the third group 18 ml of vinegar three times a day, the fourth group had to drink half a pint of seawater a day, the fifth group were given two lemons and an orange per day for six days (they then ran out of these fruits), and the last group was given a mixture of mustard, garlic and radish root which gave the sailors diarrhoea! After two weeks, James Lind reported that only two sailors were fit for work – those that had been given the lemons and the orange. This experiment was important because it has been described as the first clinical trial in medical history. Strangely, Lind was not convinced that the disease was caused by the lack of fruit in the diet – he was convinced that scurvy was caused by a disease of the digestive system! It was in 1770 that the British Admiralty decided that sailor should be given regular lemon drinks to reduce scurvy on British ships.

We now know that scurvy is caused by a deficiency of Vitamin C in the diet and that it is essential for the diet to contain foods that are rich in Vitamin C such as lemons, oranges, blackcurrants, strawberries etc.

In 1740, George Anson left the city of Portsmouth with eight ships and 1,854 men aboard and sailed around the world. When he returned four years later, only 188 of his original crew returned with him, the rest, 1,666 men, had died of scurvy.

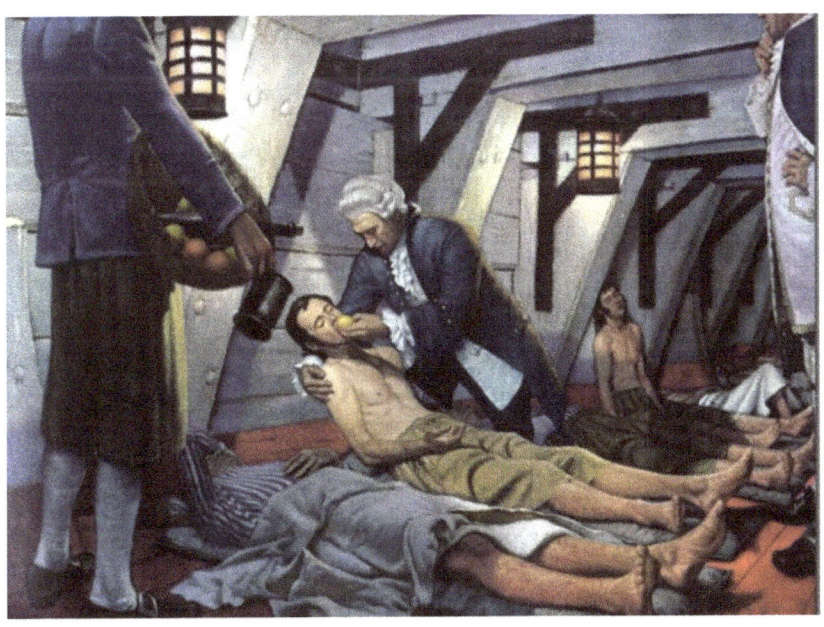

5. A painting of James Lind on board HMS Salisbury feeding one of his patients with a lemon full of Vitamin C.

6. A reconstruction of HMS Salisbury, the ship where James Lind Carried out his research into scurvy and Vitamin C. The ship was 43m in length and had 50 guns. The ship was broken up on 24th April 1761.

7. It was not only sailors who suffered from scurvy. Prospectors searching for gold in the gold rush in the Yukon, California, in 1849, also suffered from scurvy. Gold prospectors bought citrus fruits from street vendors to protect themselves against the disease.

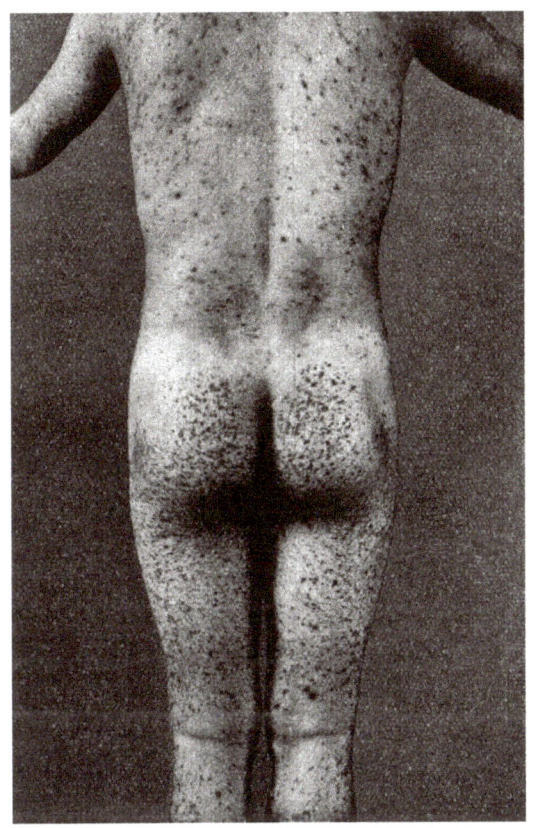

8. A person suffering from scurvy

9. This person, suffering from the corkscrew symptom of vitamin C deficiency, was given Vitamin C tablets twice a day for two weeks and the skin returned to normal.

3. How to Measure the Blood Pressure of a Horse

A horse was laid on its side and tied to the ground. A brass tube was fitted into one of the blood vessels of the neck of the horse and then a 3- metre- long glass-tube was fitted vertically into the brass tube. The blood then rose to a height of 8 feet 3 inches (248 cm). This experiment was carried out in 1733 by a brilliant English scientist called Stephen Hales, who was born in 1677 and died in 1781. He repeated this experiment on three horses, one ox and 20 dogs. He had no formal qualifications in science and was a clergyman all his working life. He realised that his experiments were cruel and gruesome, but he honestly believed that such experiments were of benefit to mankind. You must realise that attitudes towards such experiments were very different to those of today. Not only did he measure the height to which the blood rose in the glass tube, but he also observed that the level of the blood would rise and fall about four inches (10 cm) which gave a pulse. When he carried out the same experiment on a sheep, he found that the blood rose to a height of 6 feet 5 inches (193 cm). His experiments with dogs showed many differences in blood pressure depending on the breed of the dog and its size. He never measured the blood pressure in humans, and he stopped his experiments on live animals soon after 1733. Unfortunately, many of the animals used in his experiments died. However, he is known as "the father of blood pressure" because his experiments with live animals laid the foundations for an understanding of blood pressure.

Stephen Hales was also very interested in plants and he carried out experiments in photosynthesis and the loss of water through the leaves in a process called transpiration. However, he thought that his greatest contribution to science and to society was the development of ventilation apparatus. It was thought that many diseases were caused by "bad air" and the only way to improve the health of people was to remove the "bad air" from crowded places. He therefore developed bellows that would be operated by men or by windmills and these were used in overcrowded prisons, hospitals, below decks on ships and even in the Houses of Commons.

10. A painting of Stephen Hales measuring the blood pressure of a horse in an experiment of 1733

Animals	Weight of animal (kg)	Height of blood in the tube attached to blood vessel in neck (cm)	Speed of blood flow in the aorta (aorta is the main blood vessel leaving the heart) (cm/second)
horse 1	(No results)	248	4
horse 2	374	251	(No results)
sheep	41	193	7
dog 1	6	99	5
dog 2	24	81	5.5
ox	725	(No results)	1

A table showing some of the results of Stephen Hales' experiments into measuring various properties of the blood supply in different animals. Notice that he did not carry out the same experiment on all his animals. (His original units of measurements have been changed to modern units)

4. The Flesh-Eating Human Bot Fly

There really are some very nasty insects. One of the nastiest and most gruesome is the human bot fly which lives in Central and South America. The female bot fly lays its eggs on the body of a mosquito. The mosquito now lands on the skin of a human and the eggs fall from the mosquito onto the human. Now for the nasty part – the eggs hatch into larvae and they burrow into the skin and leave a hole in the skin so they can obtain oxygen. Under the skin, the larvae feed on human flesh and grow bigger and will live under the skin for up to four months, increasing in size with the infected part of the skin enlarging and becoming inflamed. Eventually, the larvae pop out of the hole and fall to the ground. Here they turn into the adult bot fly. The adult bot fly has no mouth and therefore does not eat. All it does is mate, find a mosquito, lay eggs, and then die.

There are many true stories of people feeling the bot fly larvae moving underneath their skin. The larva has many spines which dig into the flesh, and this makes it very difficult to remove. Apart from surgery, several methods have been used to remove the larva – some methods very simple indeed. The hole in the skin through which the larva uses to obtain oxygen is closed using, for example, Vaseline, nail varnish or super glue or even Sellotape. This cuts off the oxygen supply to the larva and it dies. After 12 to 14 hours, the substance or Sellotape is removed, and the area squeezed very carefully, and tweezers used to extract the dead larva. If any of the larva is left behind, then severe infection can occur.

The bot fly is common in Central and South America and many people become infected. However, they do not transmit any disease, but can cause extreme discomfort and pain. There are many species of bot fly and each of the species infects a particular animal. Some feed on horses, some on cattle, sheep, rabbits, and deer. Bot flies are also called warble flies, heel flies and gad flies. In sheep and goats, the flies lay their eggs in the nostrils of sheep and goats and can cause severe distress to the infected animals – difficulty in breathing and can even make their way to the brain. In cattle, the heel fly attaches its eggs to the feet of the cattle. They then hatch and the larvae eat their way up the legs and exit through the skin of the infected animal where they drop to the ground, hatch into adult flies and the process continues all over again.

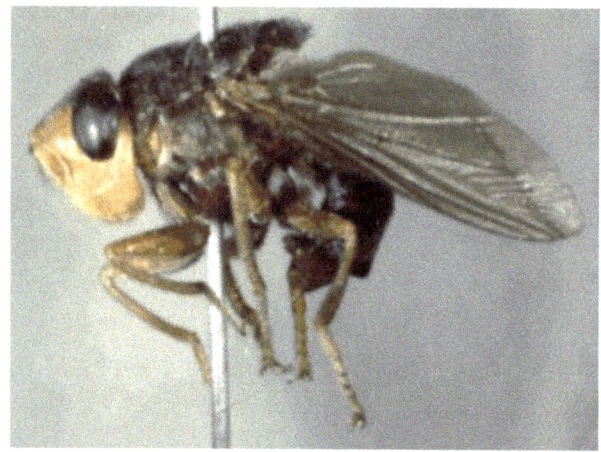

11. Human bot fly – one of the nastiest of all insects. It has no mouth and does not feed. All it does is to mate and lay eggs onto a mosquito.

12. The larvae of the human bot fly. The white "tube" is a breathing tube to take oxygen from the air to the larva. As the larva eats its way through flesh and deeper into the skin, then the tube grows longer. Note the spines which make the larvae difficult to remove.

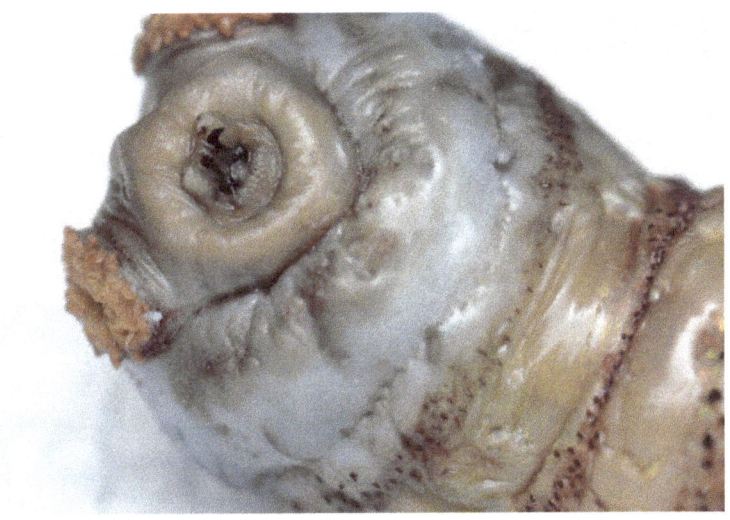

13. The head of the human bot fly larva. Notice the two black teeth which it uses to eat its way into the flesh.

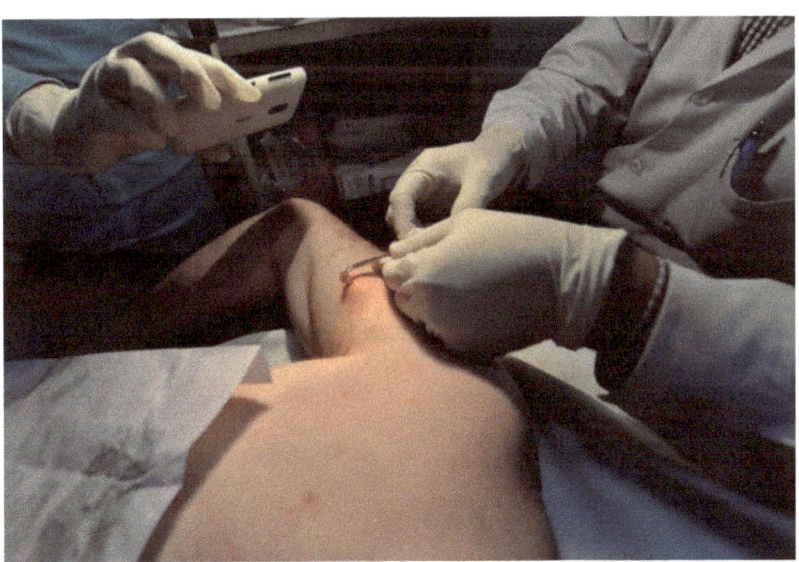

14. Removing a bot fly larva from an infected person. This has to be carried out carefully so as not to leave any of the larva inside the flesh which could rot and cause an infection.

15. A comparison between a bot fly larva and the end of a finger to give some idea as to the size of the larva.

16. Cattle can also become infected with the larvae of the fly.

5. The Most Dangerous Plant
in the Country – The Giant Hogweed

You must not think that the most dangerous living things are always animals such as snakes, crocodiles, and the great white sharks. There are some very dangerous plants and one of them is the Giant Hogweed. The plant is common in this country and is spectacular, growing up to five metres tall with large flower heads 60 cm in diameter and leaves three metres in length. Unfortunately, the plant has a chemical that is found in the roots, stems, leaves, flowers, and seeds that causes severe blistering of the skin if you touch the plant. If this chemical touches the skin and the skin is then exposed to sunlight, severe burning to the skin occurs. The chemical is called furocoumarin. Just touching the plant causes blistering. The chemical damages the skin and if the skin is then exposed to sunlight, the blistering to the skin is very serious indeed.

Because the giant hogweed is so dangerous, it is against the law to grow the plant in gardens. In the wild, the plant grows on the banks of streams and rivers, but it can also grow in drier areas. Animals such as dogs may also suffer from the effects of the plant because dogs like to run through vegetation and may come into contact with this dangerous plant.

If you have been in contact with the plant, you should wash gently the area of the skin with soap and water and remove and wash clothing to avoid further contamination. The blisters may take several weeks to heal but scarring of the skin may occur and be visible for several years. The affected area may remain sensitive to sunlight for several months.

17. Do not be fooled by this innocent white flowered plant growing along the bank of the river. It is one of the most dangerous plants to be found growing in this country and other countries.

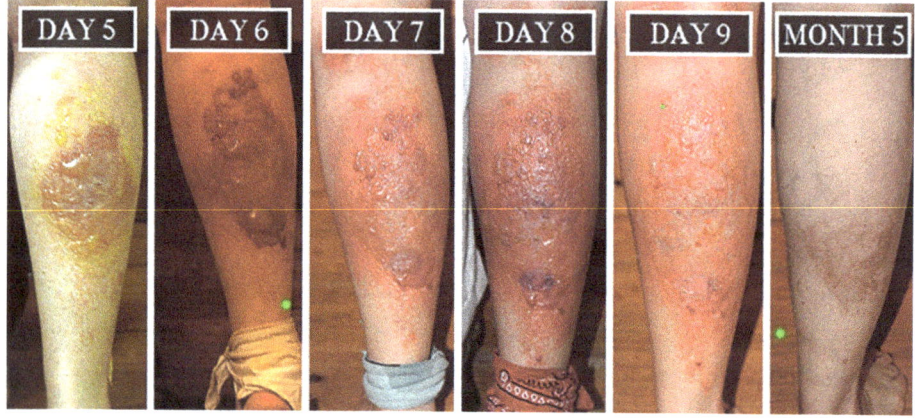

18. The sap from the Giant Hogweed can cause serious damage to the skin which can last for weeks and even months.

6. No Anaesthetic or Pain Killers

You are a young man of 25 years of age and suffering from extreme pain in the bladder. The pain is so bad that you decide to have an operation to relieve the extreme discomfort you have had for months. You remove your underclothes and trousers and sit in a chair. Your arms are tied to the arms of a chair and your legs are pushed up across your chest and then tied to your chest. Three strong men keep tight hold of you so that you cannot move. In this position, you expose the skin between the anus and the scrotum. This area of skin is called the perineum. The surgeon then makes a four-inch cut in the perineum and cuts away the flesh beneath the perineum until he reaches the wall of the bladder. He now cuts open the bladder wall. He then inserts through the cut in the perineum and bladder wall an instrument called a "Duck Bill" which he uses to grab a large stone called a bladder stone and removes the stone through the bladder wall and perineum. The size of the bladder stone which has been the cause of all the pain is about the size of a tennis ball. You were given no anaesthetic or painkillers for this terrifying and unbelievable painful operation; they hadn't yet been discovered. This operation was carried out on 26th March 1658 on one of the most famous men in history – he was called Samuel Pepys. He was famous because he wrote a diary in which he recorded in one million words his daily life for nine years. Because he lived in London, the diary records daily life in this city and he wrote about his personal life (including the operation), the weather, the theatre and social and political life of the city. He recorded details of the Great Plague of London and the Great fire of London. His dairies are one of the most important documents describing life in the London area during the second half of the 17th century. He stopped writing his diaries because of failing eyesight.

However, what happened after the removal of the bladder stone? Medical science and surgeons of the day did not believe in stitches and therefore the bladder wall and the perineum were left open. They covered the bladder wall and the perineum with a cloth that was covered in a mixture of egg yolk, rose vinegar and oils and hoped that the cuts would heal naturally. The surgeon was called Thomas Hollier and he was described as one of the best surgeons in the city, not because of his medical knowledge, but because of the speed in which he could remove bladder stones. Samuel Pepys' excruciatingly painful operation lasted only 55 seconds! There was always the risk of the cuts being infected, but he was rather lucky – he was the first patient of the day for Dr Hollier and his hands and instruments were clean! It was normal for a surgeon to use the same instruments throughout the day without cleaning!

Pepys continued to live for another 50 years. He became a member of parliament and had a library containing over 3,000 books. He died in 1703.

19. A drawing of the removal of the bladder stone from Samuel Pepys. Dr Hollier is at the front. Notice the strong men used to hold Pepys down and stop him from moving, even though he was tied to the chair.

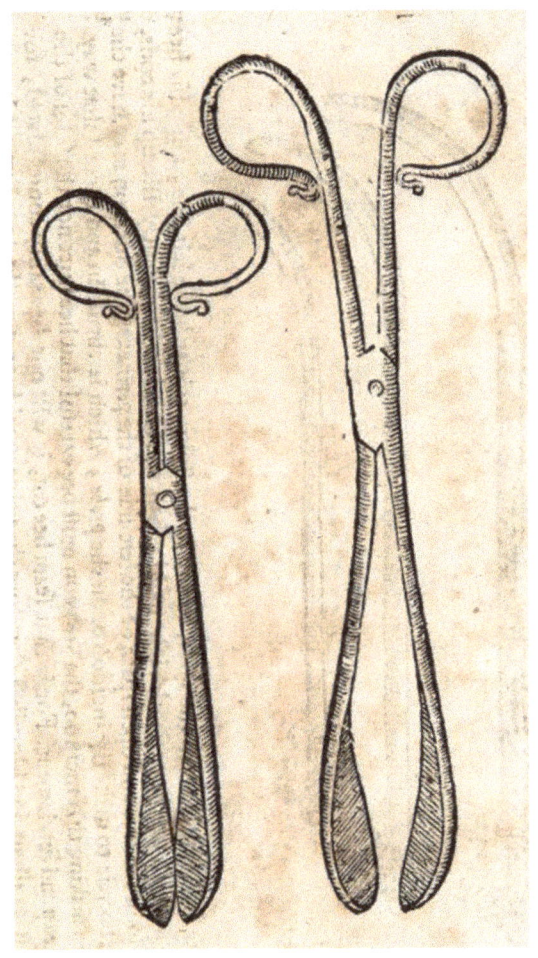

20. Duck bill forceps, similar to those shown above were used by Dr Hollier to remove the bladder stone from Samuel Pepys.

7. Alexis St. Martin –
A Human Test Tube

On the 6 June 1833, Alexis St. Martin, aged 28 years, was accidentally shot in the left side of his body by a shot gun fired from less than one metre. The wound was terrible, and it seemed there was little chance of him surviving. Dr William Beaumont, an army surgeon who lived in Michigan USA, was called to attend the seriously wounded Alexis. When he arrived, Dr Beaumont found a large hole with the contents of the stomach pouring out of the wound. There were also pieces of shattered ribs and some of the lung exposed. Dr Beaumont did the best he could and stitched the wound, but it left a hole in the side of Alexis' body, which never closed, and he died aged 78 with the hole still visible.

This terrible accident, however, was to provide medicine with an understanding of digestion in the stomach. Alexis, although badly wounded, was given food, but all that happened was the food passed down the food pipe into the stomach and then out of the hole! So, Dr Beaumont fed Alexis nutritious food through his anus! Although this kind of feeding may be disgusting to you, it kept Alexis alive and after 17 days, Dr Beaumont found that food given by mouth did not flow out of the hole but began to remain in the stomach.

Dr Beaumont now realised that he could carry out experiments in how food was digested in the stomach by inserting different kind of food through the hole and into the stomach of Alexis. He tied pieces of food with silk thread and pushed the food through the hole and after a certain time pulled the food out of the stomach using the thread and observed what had happened to the food. As you can imagine, Alexis was not keen on the idea of becoming a "human test tube". Beaumont made Alexis his servant, and over a period of nine years, carried out 200 experiments, all connected to Alexis and the hole in his stomach. He placed a tube through the hole and collected the liquid produced by the stomach which was known as gastric fluid. This fluid was added to food in test-tubes, and he observed digestion outside the stomach.

He discovered:

(1) that the gastric juice contained hydrochloric acid, which was made by the stomach wall,

(2) digestion in the stomach was due to chemicals,

(3) digestion was helped by the stomach muscles which moved the stomach and helped to mix the food with the gastric juices,

(4) vegetables were digested slowly,

(5) milk curdles in the stomach,

(6) temperature and physical activity affect digestion in the stomach.

Such were his discoveries that Dr William Beaumont is called the "Father of Gastric Physiology".

21. Painting of Dr William Beaumont draining gastric juice from the stomach of Alexis St. Martin

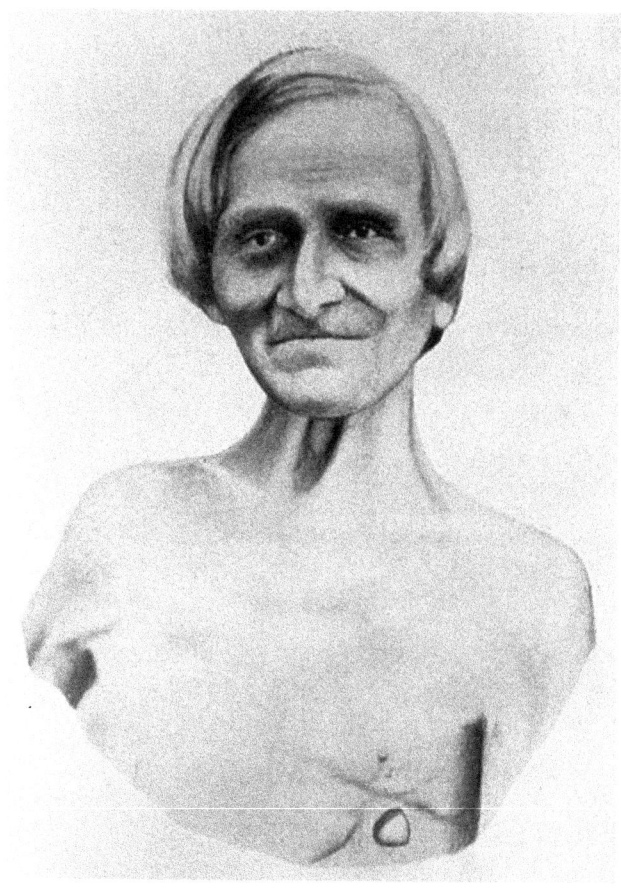

22. Drawing of Alexis St. Martin showing the hole on the left side of his body

8. The Bullet Ant

Justin Schmidt is an American entomologist (a person who studies insects) who is interested in a group of insects known as Hymenoptera, which includes 150,000 different species of bees, wasps, and ants. He is particularly interested in those that have nasty stings. He allowed himself to be stung at least 1,000 times by 150 of the most painful Hymenoptera stings. Based upon his findings, he produced the Schmidt Pain Index, which placed the insects in a kind of league table based upon the pain he experienced. The top three Hymenoptera were the Warrior wasp, the Tarantula Hawk Spider wasp and the Bullet Ant. Of these, he placed the Bullet Ant at the top of the league because the pain lasted for 24 hours – the Warrior wasp lasted two hours and the Tarantula Hawk Spider wasp five minutes.

Schmidt described the pain after being stung by the Bullet Ant as "like walking over flaming charcoal with a three-inch rusty nail embedded in your heel". The Bullet Ant injects, using its stinger, a nerve toxin which causes excruciating pain, temporary paralysis and uncontrollable shaking of the bitten area, and these symptoms remain for about 24 hours. After this period, the body returns to normal and there is no danger. There is no danger of being bitten by the ant in this country as it only lives in the jungles of South America.

The Setere-Mawe people live in the jungle of South America. When the young men of the tribe reach 14–16 years of age, they deliberately allow themselves to be stung by the Bullet Ant. A glove is made of large leaves and Bullet Ants are placed inside. The hands of the young men are then covered in charcoal, which is believed to reduce the sting of the insect! The hands are now placed in the gloves. Obviously, the Bullet Ants bite the hands, and the young men have to endure the excruciating pain. Over a period of many months or even years, this ceremony is repeated about 20 times and when this ritual is completed, the young men have now reached manhood and become warriors!

23. The Bullet Ant – the most painful sting of any insect.

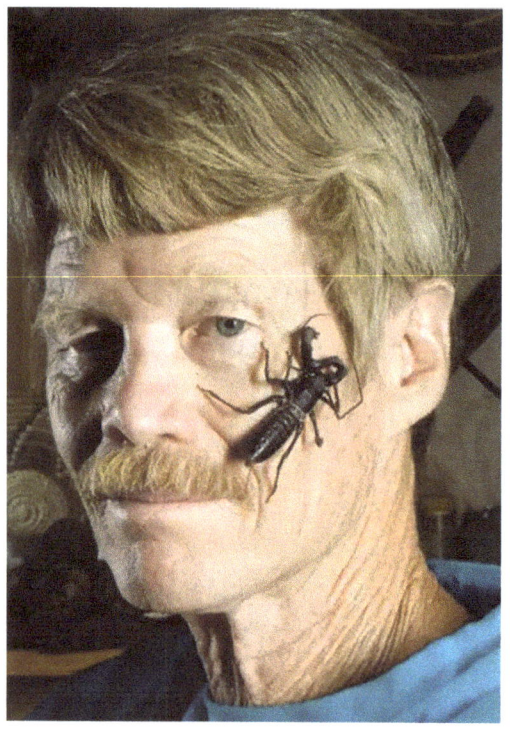

24. Justin Schmidt, the "King of Stings"
with a 2.5cm bullet ant on his face. Insect.

25. The stinger of the bullet ant.

26. The glove being prepared for the ritual.

9. Small but with a Deadly Sting – Scorpions

Arachnophobia is a fear of spiders, and there are about 45,000 different species of spider in the world, and all are poisonous, but most are not dangerous to humans. One group of spiders that frightens most people are scorpions. There are about 1,750 species of scorpion, but only about 20 have sufficient venom to kill a human. About 1,200,000 people are stung by scorpions every year, resulting in about 3,250 deaths.

The most dangerous scorpion is called the Indian Red Scorpion, which lives in India, Pakistan, Sri Lanka, and Nepal. It is 8 cm in length and, like all scorpions, is carnivorous feeding on insects, small lizards, and even small mice. Scorpions detect vibrations of movement made by their prey through their legs and wait under stones and then ambush their prey. If the prey is small, then they crush it using their large pincers. If the prey is much larger, the stinger is used for injecting venom and killing. Once having eaten, the scorpion can go for many months, even a year, without feeding. People stung by scorpions have severe pain, vomiting, sweating, heart pain and breathing problems with fluid filling their lungs.

In the deserts and drier areas of southwest America (e.g., Arizona, New Mexico, and Utah) the most dangerous scorpion is the Arizona Bark Scorpion. At 5 cm in length, it is a very fast mover and is to be found under rocks, in crevices outdoors, but also lives indoors in cupboards and drawers and it hides in shoes. It is particularly dangerous for old people and young children who often have a severe reaction to the venom.

The deathstalker scorpion has a very powerful venom and its excruciating sting is about 100 times more painful than that of a bee. However, deathstalker venom is collected by "milking" the stinger, which is a very dangerous operation indeed! The scorpion is held in one hand and then the stinger is squeezed by using forceps, held in the other hand. A very small drop of the venom is collected from the end of the stinger. Chemicals in the venom have shown promise in treating certain brain tumours and therefore large quantities are collected by "milking". Collecting one gallon of venom takes 2.6 million "milks" with a value of $39 million dollars – the most expensive liquid in the world!

27. At 9 cm in length, the Indian Red Scorpion is not the largest scorpion, but is the deadliest. It is found in India, Pakistan, and Sri Lanka. It feeds on cockroaches, lizards, and small rodents. It gives birth to live young called "scorplings". It is responsible for the death of about 70 people a year in India. The venom causes fluid to fill the lungs and causes breathing problems

28. The Emperor Scorpion is at 20cm in length, the largest scorpion in the world and is found in the rain forests of West Africa. The adults capture and crush their prey (in this case a fly) using their large pincers. The juvenile scorpion prefers to use its venomous stinger to capture its prey.

29. At 15cm in length, the Spitting Thicktail Black Scorpion is the most dangerous in South Africa. It can "spit" venom up to 1m and if it enters the eye, it can cause temporary or even permanent blindness. Also notice he large pincers for capturing prey.

30. The stinger of the Death Stalker Scorpion which lives in North Africa. Its venom is collected for medical research and has been shown to have some promise in treating certain brain tumours.

10. Sea Wasp Box Jellyfish – The Most Dangerous Sea Creature

Northern Australia has some magnificent sandy beaches, and together with clear skies, high temperatures and inviting warm seas, nothing could be more soothing than a swim. However, be warned – before you enter the sea, you may first want to dress in a stinger suit (see photo opposite) to protect you against what has been called the most dangerous sea creature – the Sea Wasp Box Jellyfish. This animal is about the size of a football and has 60 tentacles, each about three metres long, with each tentacle having about 5,000 stinging cells. It has good eyesight – it has 24 eyes! It is light blue in colour and almost transparent and this makes it almost impossible to see when swimming. Many swimmers wear stinger suits to protect themselves from the jellyfish. Normally, if a swimmer without a stinger suit only comes into contact with a small number of tentacles and although the sting is excruciating painful, the victim recovers without too much hardship. However, if a person is not wearing a stinger suit and receives a very large number of stings from many tentacles, then death can occur within two to five minutes. One Sea Wasp Box Jellyfish contains enough venom to kill 600 adults! In the previous seventy years, it is though that over 5,500 people have been killed by this sea creature with up to 40 people a year killed in the Philippines.

Each stinging cell of the jellyfish contains a small venomous dart which it fires into the skin. The dart causes very severe pain, and the venom can cause a heart attack. It is the most dangerous jellyfish in the world. Beaches have bottles of vinegar for the public to use if swimmers have mild attacks from the jellyfish. The vinegar stops the stinging cells from releasing their venom, although recent research has shown that vinegar is not as effective as was originally thought.

31. The Sea Wasp Box jellyfish is well camouflaged, and this is one of the dangers of swimming in the sea off the coast of northern Australia.

32. Protective stinger suits made from tough nylon are used by swimmers and surfers from being stung by the jellyfish.

33. Signs are posted on beaches in northern Australia warning swimmers of the dangers of the Sea Wasp Box jellyfish

34. Danger notices provide the symptoms of being stung by the Sea Wasp Box jellyfish and the treatment of the stings are to be found on many of the beaches of northern Australia.

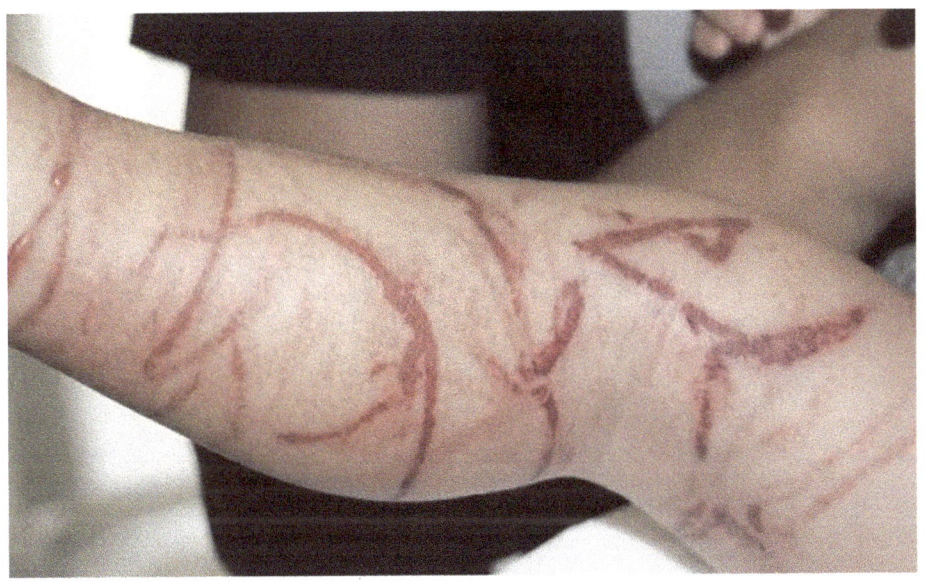

35. Stings from the tentacles of the Sea Wasp Box Jellyfish can cause serious damage to the skin

11. The Blue Ringed Octopus –
Small but Deadly

There are over 300 species of octopus, all of which are venomous, but only the Blue Ringed Octopus is deadly to humans. The Blue Ringed Octopus lives in rock pools and around coral in the Pacific and Indian Oceans. It is brightly coloured, only about 10 to 20 cm in length, weighs about 30 g, has two eyes and eight tentacles, and eats shrimps and small crabs and only lives for about two years. The octopus uses it arms covered in suckers to capture its prey. The arms are then used to pass the prey to its mouth. The octopus has a beak positioned at its entrance. The beak is the hardest part of the octopus and is shaped like the beak of a parrot and works like a pair of scissors. The beak breaks open its prey and at the same time it injects a venom called tetrodotoxin. Strangely, the Blue Ringed Octopus does not make its own venom but is made by bacteria which live in the salivary glands of the animal. The beak is very powerful and can easily penetrate skin and even a wet suit and the venom passes from the beak into the skin. The venom is 1,000 times more powerful than cyanide.

However, it rarely bites humans and only about five deaths from the venom of this octopus have been recorded. When the octopus injects only a small amount of venom, the person will suffer from paralysis, temporary blindness and produce large amounts of saliva. Each octopus carries enough venom to kill over 24 people and if enough venom is injected, a person can die within 20 minutes.

When left alone, the Blue Ringed Octopus is not blue at all. It is only when the animal feels threatened that the blue rings appear. Each animal has about 60 rings, which are normally hidden by muscle and when the animal feels threatened, the muscles relax to reveal the rings. Each blue ring flash takes about 0.3 second.

36. The octopus feels threatened, and the blue rings are beginning to appear

37. The Blue Ringed octopus feels threatened and has blue rings showing.

38. Close-up photograph of the blue rings of the octopus

39. A VERY DANGEROUS thing to do is to handle the octopus when it obviously feels very threatened. The photograph also shows the small size of this deadly animal.

12. Poison Dart Frogs

There are about 200 different species of Poison Dart frogs. All of them are small, most very colourful and have many different patterns on the skin. Although some are very poisonous, some are not dangerous at all. They live in warm, damp conditions that are found in the jungles of Central and South America. The largest is about 55 mm in length, the smallest 15 mm. They eat flies, ants, centipedes, and small beetles.

If the frogs are kept in a different environment to what they are normally used to and are given a different diet, they do not make the toxin. This explains why they can be kept as "pets" and their owners are not poisoned when they are handled. It has been suggested that the toxin is made using the toxic chemicals that are found in their normal diet. Each species makes a different poison which is stored in glands underneath the skin. In general, the more colourful the frog, then the more dangerous is its toxin. The toxin is not used to kill its prey, but instead to deter predators and it is very effective in doing so.

As far as it is known, only one animal eats Poison Dart frogs and that is the 25 cm long Fire Bellied Snake. It is not known why this snake does not die after eating these frogs.

One of the main reasons why these frogs are so well known is that the poison from their skins is used on the ends of darts, arrows and spears which are used by certain tribes in the jungle to kill animals for food. Only three species of the 200 different frogs are known to be used for his purpose, including the Golden Poison Frog.

40. At only 55mm in length, the Golden Poison Frog, is the most toxic of all the poison dart frogs. It lives in the tropical rainforests of Colombia in South America. It is said that 1g of its toxin is enough to kill 15,000 humans!

41. The Yellow banded Poison Dart frog lives in the tropical rain forests of Venezuela in South America. Its colouration is a warning to predators.
The toxin causes heart failure.

42. At about 50 mm in length, the Three Striped Poison Dart Frog is found living in the northern areas of South America. After the tadpoles have hatched, the male frog carries them on his back and deposits them in water filled hollows in trees and plants.

43. At about 20 mm in length, the Strawberry Poison Dart frog lives in Costa Rica and Panama in Central America. The adults show a high degree of parental care in looking after their tadpoles

44. The Red Poison Dart Frog is about 12 mm in length and lives in small groups of five or six in the rain forests of Peru

45. The 25cm long Fire Bellied Snake lives on a diet of poison frogs. It is not known why this snake can eat such dangerous frogs and not be harmed

13. Gympie-Gympie or the Suicide Plant

This plant was first recorded in 1866 in Queensland, Australia, when a horse brushed against the leaves of this green bush, "went mad" and died within two hours. There is also the story of an army officer when in the forests of Queensland, using the leaves of this plant as toilet paper and afterwards was in so much pain that he committed suicide by shooting himself.

Gympie-Gympie means "stinging tree" and the name was given to the plant by the Gubbi Gubbi people, an indigenous tribe of Australian aborigines. There are many different species of stinging trees and shrubs, and some are far more dangerous than others, the most dangerous being the Gympie-Gympie plant which grows to a height of three metres. The plant belongs to the nettle family of plants – the same family of plants as the stinging nettle found in this country. The leaves, stems and fruits are covered in hairs and when the plant is touched, the hairs break off and become attached to the skin. Each hair has a small bulb attached to it and when broken it releases a dangerous venom into the skin which causes a rash. The venom causes excruciating pain, prevents sleep, and lasts for several days or longer. However, bouts of pain can continue for months or even years. It took a botanist (a person who studies plants) called Ernie Rider two years to overcome the pain which happened each time he took a shower! A soldier called Cyril Bromley fell into a Gympie-Gympie plant during training in World War 2. The pain was so intense that he had to be strapped to his bed for three weeks.

It is very difficult to remove the hairs once they have penetrated the skin. One of the most common methods is to apply wax hair removal strips and then pull them off and this removes the hairs from the skin. You do not even have to touch the plant to experience the painful effects of the venom. The hairs can be in the air near to the plant and when breathed in, can cause swelling of the mouth and tongue, bleeding nose, throat irritation and breathing difficulties.

The Gympie-Gympie plant can cause problems for foresters, land surveyors and construction workers as well as livestock, campers, and travellers. Warning notices warn visitors about the dangers of the plant.

Strangely, there are one or two animals that can eat the leaves of the plant and not be affected. Some beetles eat the leaves of the plant and the Red Legged Padamelon, a small animal related to the kangaroo, has no difficulties in eating the plant.

46. It is difficult to believe that these 'innocent' looking leaves belong to one of the nastiest plants on earth.

47. A photograph taken through a microscope of the hairs from a leaf of the Gympie-Gympie plant. Each hair has a small bulb at the end. When the hairs and bulbs are broken the bulbs releases a very nasty chemical producing "the worst pain you'll suffer in all your life", and leaves behind no cuts or scratches on the skin.

48. When dealing with the Gympie-Gympie plant, protective clothing must be worn.

BEWARE STINGING TREE

VISITORS ARE ADVISED TO BEWARE OF THE STINGING TREE. CONTACT CAN CAUSE SEVERE PAIN AND DISTRESS. IF STUNG – SEEK IMMEDIATE MEDICAL ATTENTION ..... DIAL 000

危険 ステインギグーツリーにご注意！

この辺りには、スティングツリーがございます。トゲに刺されると激痛を引き起こしますのでご注意下さい。万一、刺された場合は直ぐに治療を受けてください。

救急電話番号 000

49. Warning signs are posted if the Gympie-Gympie plant is in the area

50. The Red Legged Padamelon can eat the leaves of the plant without being harmed.

14. Mawson and Vitamin A

You have probably never heard of Douglas Mawson. Born in England, he moved to Australia when an infant and he lived the rest of his life in Australia. He was destined to become a great Antarctic explorer and is remembered for his unbelievable survival on the coldest continent on earth as "the greatest story of lone survival in polar exploration". Mawson was a member of the Australian Antarctic Expedition of 1912 whose aim was to map areas of the Antarctic. Mawson and his two colleagues left their Antarctic base on 10 November 1912. Mawson and his two colleagues, Belgrave Ninnis and Xavier Mertz, were pulling two dog sledges across the Antarctic ice back to their expedition base when Ninnis and his sledge and dogs fell 150 feet down into a crevasse and were killed. This sledge carried most of their food as well as dog food. This left Mawson and Mertz with one sledge and six husky dogs that were very weak from pulling heavy sledges across difficult terrain and against very cold winds. They had food for only one week and no dog food, and they took the decision that the only way they could survive was to eat the dogs. This was a particular difficult decision for Mertz, because he was a vegetarian, but he accepted that the only way he could survive was to eat dog meat. They had no change of clothing and struggled on in wet clothes and slept in wet sleeping bags. However, on 9 January 1913, with his health deteriorating, suffering from violent diarrhoea and bout of deliriousness, Mertz died. It has been suggested that he died because his body could not cope with the change from his vegetarian diet to one of meat. Mawson was now left with a journey of 160 miles across the most inhospitable land on Earth on his own. The dogs had been close to death and when killed and eaten, provided very little nourishment, although the livers were the most edible part of the animals. Mawson's lips and nose were broken open because of the low temperatures and strong winds. He had open sores in his groin because he was wearing wet clothes. To reduce the weight of his sledge, he cut it in half using his penknife on 11 January 1913. He attached his harness to the sledge, but after only few miles, his feet were so painful that he could not walk. He removed his boots and socks and found to his horror that all the skin had come away from the soles of his feet. He covered his feet with lanolin ointment and bound the loose skin back to his feet. He recorded in his book "my whole body is apparently rotting from the want of proper nourishment – frost bitten fingers, festering of my nose, salivary glands not producing saliva, skin coming off the whole of my body". He struggled on in the terrifying conditions of the Antarctic and on 8 February 1913, he reached the base camp only to see his supply ship disappearing over the horizon! However, all was not lost. Some of his colleagues from the expedition remained behind to search for Ninnis, Mertz and Mawson.

51. The last photograph of Mawson, Ninnis and Mertz leaving their Expedition Main Base on 10 November with their two dog sledges.

52. The sledge that Mawson made by cutting his original sledge in half using his pen knife.

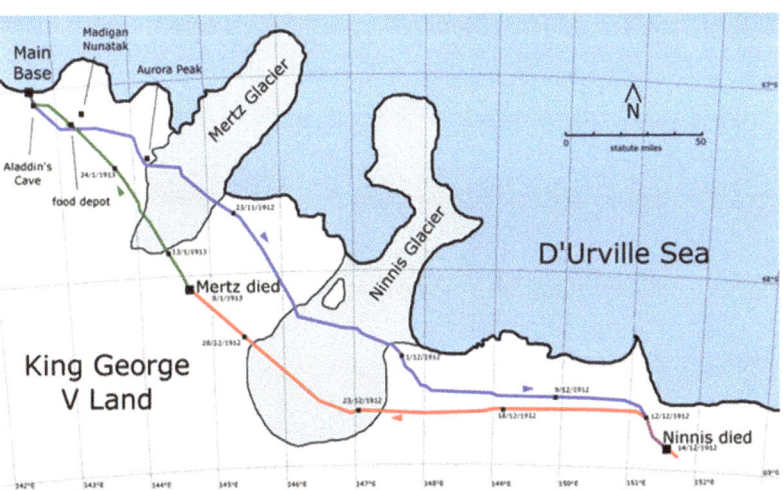

53. Map showing the route take by Mawson and his two colleagues, Ninnis and Mertz who had Antarctic glaciers named after them.

54. Mawson after his 160 mile walk alone across the most inhospitable land in the world.

15. Devil's Cherries and Beautiful Lady

One of the most poisonous plants in Europe is called Deadly Nightshade and it is found growing in this country, particularly in southern England. The plant has several different names, including Devil's Cherries and Belladonna. The plant is attractive and reaches about 1.5 metres in height, with purple flowers and very shiny sweet tasting black berries that look very much like normal cherry fruit and look attractive to children. However, all parts of the plant, including the berries, contain a poison known as atropine – hence the name Devil's Cherries. Eating two to four berries is sufficient to kill a child and an adult would have to eat up to 20 berries for death to occur. Eating any part of the plant can cause headaches, blurred vision, slurred speech, rapid heartbeat and if too much is eaten, then death will occur. Deadly Nightshade belongs to the same group of plants as the potato, tomato, and aubergine.

Although Deadly Nightshade is highly poisonous, it is used in medicine. Sometimes when an eye doctor wants to look inside your eye and needs the pupil of the eye to be wide open, he will place a very small amount of atropine on the surface of the eye and the pupil will open wide (see photograph opposite). The atropine used by the doctor is very much diluted – one drop of atropine is added to 130,000 drops of water!

An alternative name for Deadly Nightshade is Belladonna, which means "beautiful lady" in Italian. It has been used in cosmetics for over 2,000 years and was certainly used by Cleopatra who was Queen of Egypt between 51–30 BC and used by Italian women during the 14th to 16th centuries to 'enlarge' the eyes and so make themselves more attractive.

Roman archers made a paste containing atropine into which they dipped their arrows and used them to kill their foes.

Many soldiers in World War 2 received severe eye injuries during battle and atropine had to be used by surgeons to examine the eyes. During the War, 200 tons of leaves and roots of the plant were collected for the extraction of atropine.

Atropine is commonly used to neutralise the effects of poisonous nerve gases that are still used in conflicts in some parts of the world.

55. The Deadly Nightshade plant which contains a poison known as atropine

56. Eating just 2–4 of these juicy looking fruits from the Deadly Nightshade plant is sufficient to kill you.

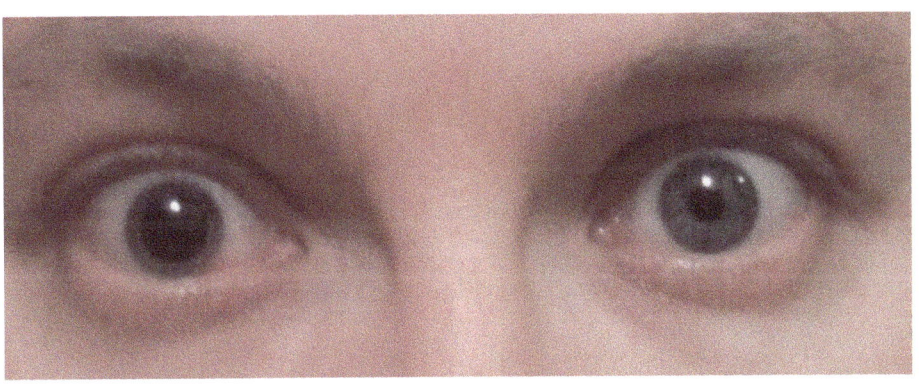

57. A drop of atropine solution has been added to the right eye. Notice the dilation (enlargement) of the pupil of right eye and compare this to the normal size of the left eye pupil.

16. Plants That Eat Animals – Giant Malaysian Pitcher Plant

There are about 375,000 different species of plant in the world, and more are being discovered every year. In 2015, over 2,000 new plant species were discovered. There are about 750 plant species that eat animals – these are known as carnivorous plants. The most famous of all carnivorous plants is the Giant Malaysian Pitcher plant also known as the "King of Pitcher Plants" found in the wet grassland areas of Borneo. This Pitcher plant grows along the ground which has very little nutrients and it is because of this that the plant has developed these giant pitchers to capture animals, digest them and so obtain the nutrients that it requires to survive.

Each pitcher is very large, about 40 cm in height and 20 cm in width, and it holds about three litres of digestive juices. Each pitcher has a lid called an operculum which stops rainwater from entering the pitcher and so diluting the digestive juices. The pitcher is brightly coloured and sweet smelling to attract insects and other animals but also has sticky hairs to stop them escaping. Although it is mainly insects that become trapped in the pitcher, much larger animals are known to be slowly digested such as lizards and even rats! The plant is not dangerous to humans.

Even more strange is that some animals live inside some pitcher plants and come to no harm. The black spotted sticky frog, one of the world's tiniest frogs at only 10 mm to 12 mm in length, lays its frog spawn on the side of them pitcher and the tadpoles live in the liquid and are not digested because they have a sticky layer on their skin which protects them from digestion.

The pitcher plant also attracts mountain tree shrews to its pitchers by producing large amounts of sweet sugary nectar on their operculum. The tree shrew climbs onto the pitcher to reach the nectar, and in doing so, its backside is positioned over the mouth of the pitcher. The poo from the shrew drops into the pitcher and the plant digests it and so obtains nourishment.

58. The pitcher of the Giant Malaysian Pitcher plant. The pitcher is about 40 cm in length. The plant grows along the ground which has very little nutrients in the soil.

59. The pitchers can be quite big!

60. A lizard has become trapped in the pitcher and will now be digested.

61. The Mountain Tree Shrew feeds on sugar which is made by the operculum and then deposits its poo inside the pitcher plant of the Giant Malaysian Pitcher plant which the plant digests and so obtains its nutrients

62. Some frogs can live inside the pitcher without being digested.

17. Plants That Eat Animals –
Venus Fly Trap

Carnivorous plants have developed several different methods of capturing their prey. The Pitcher plants (see Chapter 16) uses a "pit fall trap". One of the most interesting and nasty carnivorous plants is the Venus Fly Trap that uses a method of capturing its prey known as the "snap trap". This plant lives in wet boggy areas on the eastern coast of the United States and, like all carnivorous plants, lives in area where there is very little nourishment in the soil. It is a small plant about 15 cm to 30 cm in height, but don't let that fool you – it is deadly in catching its prey.

Each trap is made from two hinged leaves, both of which have spiny teeth, with the inside of each leaf having three "trigger hairs". The prey is attracted to the trap by sweet smelling nectar produced by the leaves, and when the prey land on the trigger hairs, the leaves close and the prey is trapped. The leaves snap shut in about a tenth of a second, and it takes about 30 minutes for the trap to tighten its leaves and crush its prey. The live prey, consisting mainly of insects is now digested alive which takes between four and ten days, depending on the size and nature of the prey. When fully grown, each Venus Fly Trap has between five and ten traps at any one time. If the prey is too small (e.g., a very small insect) then they can escape through the spaces between the spiny teeth of the leaf. The plant can also distinguish between raindrops and its prey.

Biologists have discovered that the prey is approximately 33% ants, 30 % spiders, 10% beetle and 17% grasshoppers, with other prey consisting of caterpillars, wasps, and flies.

The plant is an endangered species. In a survey carried out in North and South Carolina (in USA) in 1979, biologists counted 4,500,000 plants. Thirty years later, only 163,951 plants were counted. The reasons for this decline included destruction of their habitat for building, road construction, and agriculture. Another important reason was the deliberate act of collecting the plants for sale. In 2019, a man was jailed for 17 months in North Carolina for collecting 970 Venus Fly Trap plants and hoping to sell them for profit.

The Venus Fly Trap is often found for sale in garden centres even in the UK, and with a little care, can easily be grown inside the home. These plants have been grown in specialist greenhouses and have not been poached from U.S.A.

63.The Venus Fly Trap is a carnivorous plant. Charles Darwin (1809–1882) the famous naturalist and biologist described it as "one of the most wonderful plants in the world".

64. Each Venus Fly Trap leaf has three of four black hairs on the inside of the leaf. When these are triggered (touched) by the movement of the prey the two leaves close and trap its prey.

65. Little does the fly know that it is about to become trapped and digested.

66. A Hover fly has been trapped by the Venus Fly Trap. The fly will now be digested alive over a period of several days.

18. Plants That Eat Animals –
The Bladderwort –
The Quickest Moving Carnivorous Plant

Bladderwort is a carnivorous plant that lives in ponds that have very few nutrients. It lives in the United Kingdom and is particularly common in the west of Scotland and the Norfolk Broads but can be found in many ponds. Most of the plant is found floating or just underneath the surface of water and is only seen when the yellow flowers appear on a stalk about 20 cm above the water. The plants trap and digest very small aquatic animals to obtain the nutrients that are necessary for survival. The plant has small leaves, about 1 cm to 2 cm in length, and has long underwater stems but no roots. Although the flowers are a beautiful yellow colour, don't let this fool you. This small plant has one of the most sophisticated methods of all carnivorous plants to capture its prey. There are over two hundred different species of bladderwort, and all use the same method to capture their prey – the bladder trap. Each plant has many underwater bladders varying in size from 1 mm to 12 mm, depending upon the species. Each bladder has a trap door which is sealed by a sticky mucus produced by the cells of the bladder. Water is pumped from inside the bladder through their walls to the outside. This means that the pressure inside the bladder is always lower than the pressure of the surrounding water. Each bladder has trigger hairs around its entrance. When the trigger hairs are touched by very small animals, the trap door of the bladder opens, and the prey is sucked inside. The bladderwort feeds on small animals such as water fleas, but species with larger bladders can suck in tadpoles.

The opening and closing of the trap door is one of the fastest movements to be found in plants. It takes about 1 millisecond (one thousandth of a second) for the bladder to open and 2.5 millisecond to close. Special cells lining the inside of the bladder release chemicals which digest the unfortunate prey which takes about 30 minutes.

67. Although the Bladderwort has beautiful yellow flowers above the water surface, it is below the surface that the plant is at its most gruesome

68. The Bladderwort has many bladders attached to the stem of the plant, which has no roots. The most gruesome part of the Bladderwort – the bladders are found underneath the water.

69. Two bladders of the Bladderwort plant seen through the microscope.

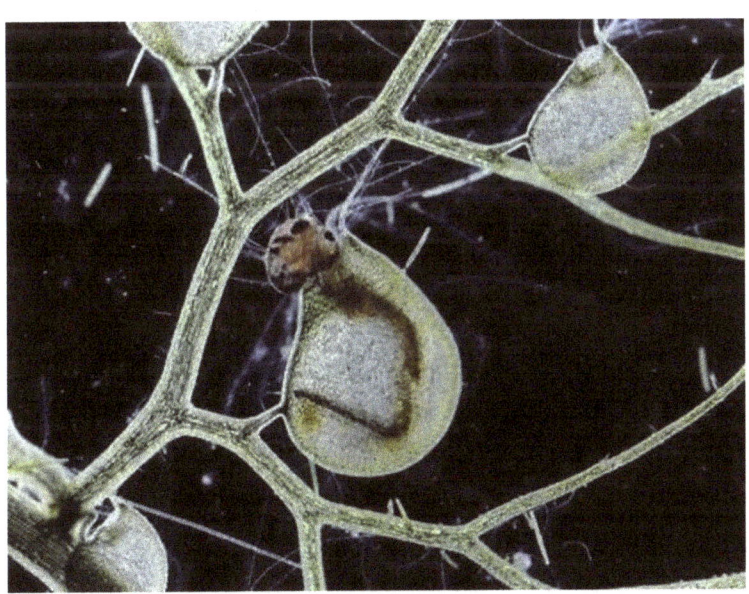

70. The bladder has sucked in the "tail end" of a mosquito larva. The head of the larva is too big to pass through the trapdoor of the bladder. The bladder will now only digest the "tail end" of the larva.

19. Plants That Eat Animals –Sundew

There are over 150 different species of Sundew, half of which grow in Australia. The Sundew that can be found in the United Kingdom is only a few centimetres high, but there is one species in Australia that grows to a height of three metres. In the United Kingdom, the Sundew is found growing in wet boggy area of moorland in Scotland, northern England, Wales, Southwest England, and Ireland. These plants grow in areas where there is very little nourishment, and they have to capture small insects and spiders in order for it to survive. It uses a "flypaper trap" to capture its prey.

Each colourful leaf of the Sundew has long hairs (which are scientifically known as tentacles), each with a drop of sticky, sweet smelling liquid at the end. Each drop is so sticky, it can be stretched to one metre in length! The smell of the liquid attracts their prey, which is trapped in the sticky liquid and cannot escape. The leaf then folds, and the tentacles secrete a liquid that digests the prey and so the Sundew obtains its nourishment. It takes about 15 minutes for the prey to die and once the prey has been digested, the leaf unfolds and is ready to capture more.

One of the first scientists to research Sundews was the famous biologist, Charles Darwin. In 1860, he wrote "at the present time, I care more about Sundew than the origin of all the species in the world". He was the first person to realise that some plants were carnivorous.

71. In the United Kingdom, Sundew is found growing in wet acid bogs of moorland in Scotland, Lake District, Wales, Ireland.

72. The Sundew leaves are colourful and attract insects.

73. Each tentacle of the leaf has a drop of sweet smelling, very sticky liquid which traps prey, and makes escape impossible.

74. Once prey has been trapped, the leaf folds over and crushes the prey which takes about 15 minutes to die. The leaf then produces digestive chemicals, and the digestion process may take several weeks. After the prey has been digested, the leaf unfolds ready to capture another unfortunate animal.

20. Parasites

Have you or your pet dog or cat had fleas? Fleas are insects that bite through the skin and then feed on blood. The flea is an example of a parasite – an organism that lives on or inside another animal that is called the host. The flea is an example of an external parasite because it feeds on the outside of its host. Some parasites are harmless, others are serious pests, and some are very horrible indeed. All animals, including humans, have parasites. There about 300 parasites that can live on or inside your body, but in this country, it is rare for any parasite to have a really bad effect on your health. It is thought that there are tens of thousands of different species of parasites. Not only do parasites live on or in animal hosts, but also on or in plants. In the United Kingdom, about 8,750 different insects are parasites with animals as hosts, and the same number of insects with plants as hosts.

In the United Kingdom, the parasites most likely to affect you are thread worms, ticks (very small spiders), fleas, bed bugs, head lice and scabies mite.

In other parts of the world, particularly in hot tropical areas, parasites are a very serious threat to a healthy life. Parasite such as tapeworms, flukes and hookworms infect different organs in the body and affect millions of people.

On the opposite page are photographs of some of the commonest parasites found in this county, with more information about these parasites in Chapter 21. In Chapter 22, you can read about some of the most horrible parasites that infect humans.

75. Human flea, 3 mm in length. They are thin so that they can move easily through hair. There are claws at the end of the legs for attaching to hairs. They can jump long distances – up to 20cm and can leap vertically 150 times their height.

76. Head louse, 2–3 mm in length. They have large claws for attachment to hair and they can move through hair very quickly but cannot jump. The red abdomen shows it has been feeding on blood

77. Lime disease tick 3–5 mm in length. They are dangerous because they can transmit through biting several diseases, particularly in the USA.

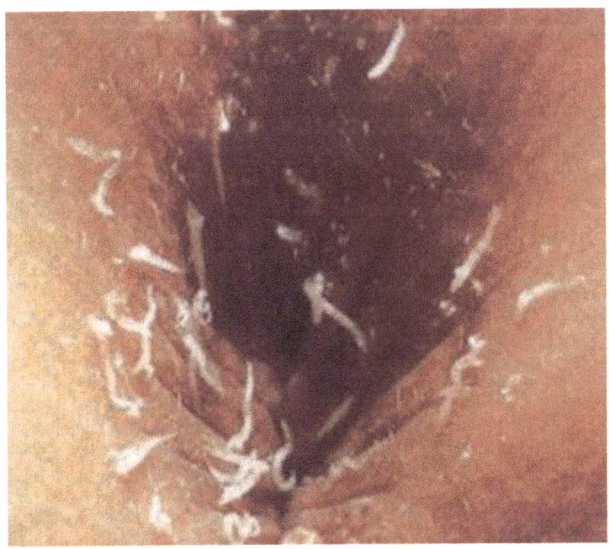

78. Thread worms (sometimes called pin worms), 8–13 mm in length, living inside the large intestine. They live inside the intestine for five to six weeks and during this time the female lays about 11,000 eggs

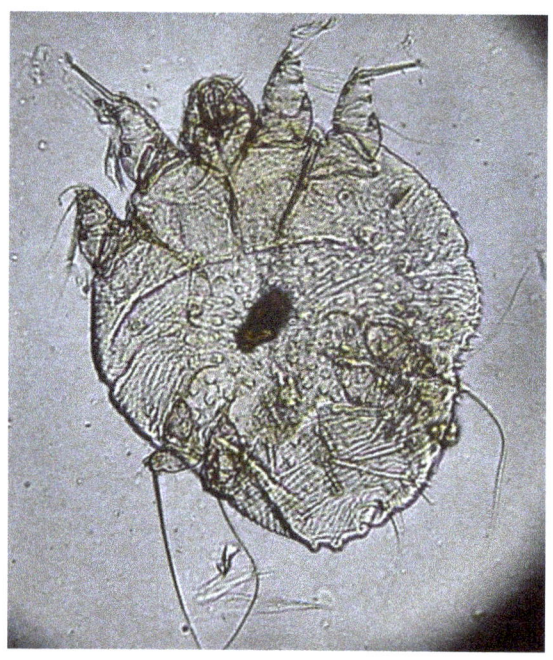

79. This horrible looking Scabies Mite (a small spider) is only 0.3 mm in length, but it infects about 100 million people.

80. Bed bug, 5 mm in length. Note how fat and enlarged is its abdomen because it is feeding on blood!
Centers for Disease Control and Prevention's Public Health Image Library (PHIL), with identification number #9820.

21. Parasites in the United Kingdom

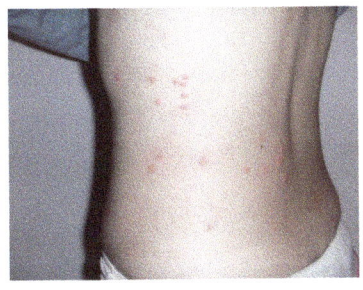

These bites are caused by the blood sucking external parasite – the flea. These small insects can be very common on dogs and cats, and it is the cat flea which is most likely to bite humans. The bites can be very irritating and cause pain and the adult flea can live for up to a year, laying about 50 eggs per day. Fleas are flattened sideways which enables them to move through the host's hair more easily. Their hind legs are long, and they can jump long distances up to 50 times their body length.

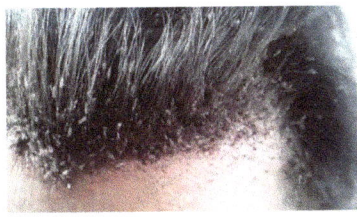

Have you ever had nits? If so, then it is the head louse that has been responsible for laying their eggs (nits) in your hair. Each egg is "glued" to a hair by the female head louse, and this means they are difficult to remove. This blood sucking external parasite cannot fly or jump, but simply walks from one person to another person's hair when they come into close contact with each other. Head lice and nits are particularly common in children between four and twelve years of age, but older people can become infected. You must look very carefully to see the nits in the photograph above. They look a little like pieces of dandruff but do not fall off the hair.

Lyme disease is an infection caused by a bacterium which is transferred to humans when they are bitten by a small spider called a tick. In 2012, a total of 7,738 people was infected with Lyme disease in the United Kingdom. The tick lives mainly on deer and when they fall off the animal onto vegetation they can be transferred to people when they brush against vegetation. Although the tick is

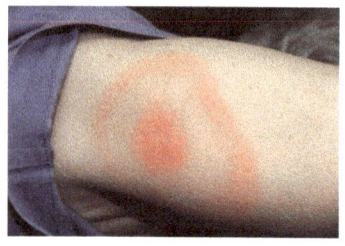

very common, only a small number of ticks carry the bacteria, but unfortunately people who are bitten can become very ill with a high temperature, headaches, tiredness, and muscle pain, with these symptoms lasting weeks or even months. A bullseye rash often appears around the site of the bite as seen in the photograph above.

Have you ever had an itchy bum? If you have then you were probably

infected with a parasitic worm called a threadworm which lives in your large intestine. The worms can be seen in your poo and appear as thin threads of white cotton about 1 cm in length. At night the female worm comes out of the anus and lays its eggs on the skin. This causes irritation around the anus and the infected person scratches and some of the eggs becomes attached to the fingers and fingernails. Children suck their fingers, and the eggs are swallowed, and the infection starts all over again. Eggs also fall onto bed sheet and clothing and can be even carried into the air where they can be swallowed. Cleanliness of the body and all clothing must be washed to stop the infection. It is the commonest worm parasite in children in the United Kingdom.

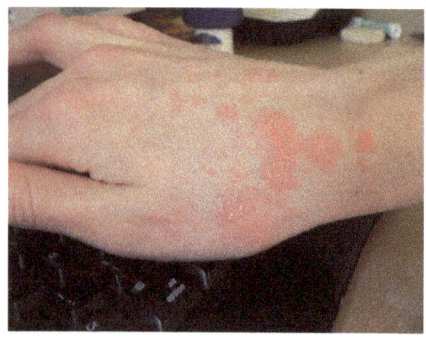

In the United Kingdom, about one person in every thousand see their doctor every month because they have a parasitic infection caused by a very small spider called a mite that burrows beneath the skin and causes very intense itching especially at night. This infection is called scabies. A person who has scabies has about ten of these mites burrowing underneath the skin at the same time. Scabies is very infectious and is passed from one person to another through skin-to-skin contact. It is common in children, but adults can also become infected. Absolute cleanliness is essential to stop scabies from spreading and clothes and bed clothes must be washed frequently. Skin creams are available to kill the mites.

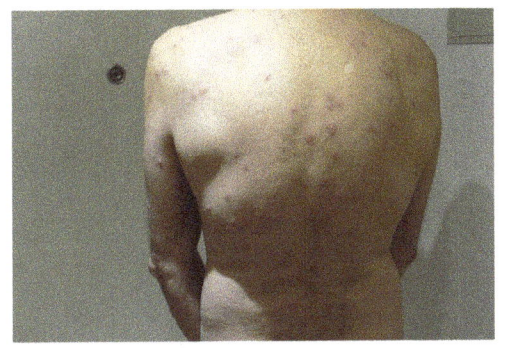

Bed bugs hide on curtains, carpets, mattresses, and furniture particularly in bedrooms. A night they crawl onto beds and onto the skin of a sleeping person where they bite into the skin and feed on blood. When the bug bites they inject a painkiller so that the person does not feel the bite. They also inject a chemical that stops the blood from clotting. The bugs hitchhike on clothes, old furniture, even onto buses, trains and aircraft. In 2014, one of the boroughs of London reported 1800 cases of bed bug infections. Thorough repeated vacuuming and the washing of clothes and bed clothes reduces the bug infections.

22. Horrible Parasites

Although there are parasites that affect humans in the United Kingdom, there are few people who have serious health problems because of parasitic infections, e.g., fleas cause nasty bites and much irritation, but they are not going to kill you. However, in some parts of the world, particularly in Africa, Asia and South and Central America, some parasites are a very serious problem for millions of adults and children. Scientists have developed medicines that control parasitic infections, but there are large numbers of people who cannot obtain medical care. Although the parasites shown here are horrendous, the worst parasite in the world is the one that causes malaria.

In 2021 a man in Thailand complained of pains in his stomach and had lots of wind in his guts. A sample of his poo was taken and when it was looked at under the microscope it was found to have the eggs of a tapeworm in it. He was given de-worming medication at bedtime. In the morning, a tapeworm measuring 18 metres in length came out of his bottom. (Why not cut a piece of string the same length just to see how long it really was). This tapeworm is shown in the photograph on the left. Strangely, most people who have tapeworms have no symptoms or only mild symptoms of infection. Humans become infected by eating raw or undercooked beef and pork which contains eggs of the tapeworm. Once inside the human the eggs develop into the adult tapeworms which can live inside the body for years and gradually grow longer and longer. At any given time about 20 million people have tapeworms.

Mango flies (sometimes called deer flies) live in the rain forests and swampy areas of West Africa and are responsible for the spread of the African eye worm which can infect humans living in these areas. If the flies feed on the blood from an infected person or from an infected animal e.g., buffalo or deer, and the flies

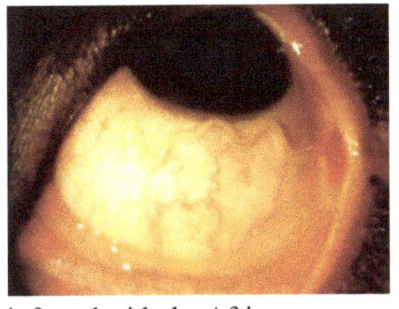

bite another person the eggs of the parasite will be transferred. Once inside the person, the eggs take about five months to develop into worms about 40 mm in length. The worms now travel around the body and even across the eyes (see the photograph above). In West Africa about 12 million people are infected with the African eye worm.

In 2019, the WHO stated that 120 million people were infected with a disease called elephantiasis of whom 40 million were seriously disfigured. The disease is carried by mosquitoes and is to be found in India, Africa, and Asia. When the mosquito bites and feeds on blood from a person with the disease part of the life history of a parasitic worm passes from that person into the mosquito. When the mosquito feeds again the disease is transferred to another person. The worms now block blood vessels and liquids accumulate in the legs. Between the years 2000 and 2019, the WHO gave 7.7 billion treatments to 910 million people to considerably reduce the transmission of the disease.

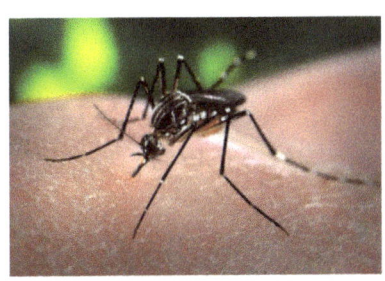

It is difficult to believe that this insect is responsible for killing more people than any other animal. Mosquitoes carry many different diseases (e.g.., yellow fever, dengue, trachoma) but the most dangerous is malaria. The mosquito does not kill but transmits the disease by biting someone who is infected with malaria and then transfers it to the next person it feeds on. Strangely it is only the female mosquito that carries the disease – it requires the protein from blood to help the female mosquito develop her eggs. The actual disease is caused by a one celled animal called a Protista which enters the blood of the person and destroys their red blood cells with devastating effects in the disease called malaria. Because it is such a serious disease, the WHO produces a report every year on malaria. In 2020 there were 241 million cases of malaria with 627,000 deaths. About 5 billion dollars were spent in 2021 in the control and treatment of malaria.

23. Fungi – Very Useful – Very Nasty

Most people would think fungi are not particularly interesting or important. With over 100,000 different species, most are microscopic in size. It is thought that there may be as many as two or three million species of fungi. Of these, 8,000 species cause diseases in plants and about 300 species can affect humans. Some species are very useful indeed, some are poisonous, some cause horrible diseases and many are harmless. Some fungi are very well known – mushrooms as food, yeasts for making bread, wine, beer, cheese, and penicillin as an antibiotic. In 2019, 1882 new species of fungi were discovered.

Fungi are to be found all over the world – in soil, in deserts, in fresh water and in sea water. Most can only be seen when they are "fruiting" – the mushroom you eat is the fruiting body of a fungus. They play an essential role in breaking down or decomposing animals and plants when they die. Fungi also help plants and in particular trees to grow. They attach themselves to the roots of trees and help to absorb minerals which are essential for the tree to grow. However, some fungi cause serious very problems especially for agriculture.

91. This rice plant is infected with the most devastating agricultural fungal disease in the world – rice blast disease. It is difficult to believe that this fungus is responsible for the destruction of 30% the world's rice supply at a cost of 60 billion dollars every year.

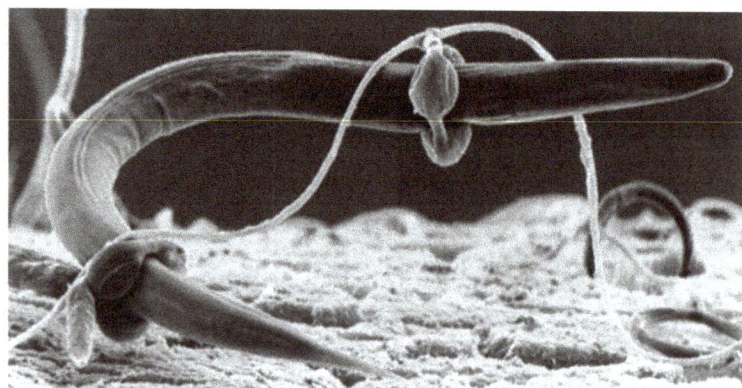

92. A spadeful of soil can contain as many as 70,000 microscopic worms called nematodes. These carnivorous worms are found in every kind of soil and there are over 160 species. Some soil fungi eat these worms. The fungus produces very fine "hairs" called hyphae which form loops (called fungal rings). When the nematode worm wriggles through the loop the hyphae tighten, and the worm is trapped. The fungus now digests the worm

One of the most horrible fungi is called Cordyceps. It is a parasitic fungus and infects living spiders and insects. Once the fungus has landed on the unfortunate animal, it grows through the skin and into the animal and then the fruiting bodies grow out of the dying animal. There are about 600 different species of Cordyceps, and each infects a different animal. Some of these fungi are to be found in the United Kingdom, but they are more common in humid tropical forests. Strangely Cordyceps fungi are used extensively in Chinese medicines where it is believed that it boosts the energy levels of the body and protects against diseases such as diabetes and asthma, but there is no strong scientific evidence for this.

93. Cordyceps fungus growing out of a caterpillar

94. A grasshopper killed by a Cordyceps fungus. The tall whitish structures are the fruiting bodies of the fungus.

95. An ant killed by a Cordyceps fungus. The large white structures are the fruiting bodies of the fungus.

24. The Doomsday Fungus

This fungus can be described as the deadliest organism on earth. This fungal disease has caused the extinction of 90 species of amphibian (frogs and toads). It is thought that the populations of a further 500 species are in decline because of this devastating disease. The fungus attacks the skin of the amphibian which then peels away from the body, a heart attack occurs, and the animal dies. At present, the fungus is found in 60 countries with the worst affected being Australia and countries in Central and South America.

The proper name for the doomsday fungus has an almost an unpronounceable name – Batrachochytrium dendrobatidis or simply for short Bd. The fungus attacks frogs and toads which live mainly in water; therefore, the fungus is easily spread.

Because the fungus causes so many amphibian species to become extinct, scientists carry out research to try and rescue species that are very close to extinction. One of these is the Mountain Chicken Frog, sometimes called the Giant Ditch Frog. It is found on only two Caribbean islands – Montserrat and Dominica. In the year 2016 only two frogs were known to be living on the island of Montserrat and on the island of Dominica only 100 of the frogs were recorded. The reduction of the population size was due to the Doomsday Fungus. Before the fungus reached the islands, tens of thousands of these frogs were to be found. The frog is the national dish of Dominica, and the hunting of these frogs is now banned on the islands with the full support of the islanders. The Mountain Chicken Frog is one of the largest in the world, weighing about one kilogram and reaching 22 cm in length, and tastes of – you've guessed it – chicken. Unlike most other amphibians, it does not breed in water, but in burrows about 50 cm deep. The frogs have been collected, bred in captivity, and then returned to their natural environment on the islands. Sometimes they are washed in an anti-fungal solution for about five minutes to deter the fungus from attacking the frog.

Research has shown that the attempts to save the frog from extinction have been a little successful.

The Mountain Chicken Frog is a giant among frogs weighing one kilogram and reaching 22 cm in length. After a devastating decline in the population due to the Doomsday Fungus, scientists have started a conservation programme with some success and numbers are now slowly increasing.

This frog is infected with the Doomsday Fungus a parasitic fungus that has caused the extinction of at least 90 different species of amphibian. The fungus causing the disease was only identified in 1998 and it is though that this fungus is responsible for the extinction and decline of at least 500 different species of amphibian.

These frogs have all died from the Doomsday Fungus and because of the devastating effects it has on amphibian populations, it has been described as "the deadliest pathogen known to Man".

Websites

https://www.ncbi.nlm.nih.gov/pmc/articles/PMC4320518/figure/Fig1/

https://www.lommelegen.no/images/74449195.jpg?imageId=74449195&width
=640&height=384&co

https://pixels.com/featured/2-demodex-folliculorum-eye-of-science.html

https://premiumvisionsc.com//wp-content/uploads/2015/04/demodex5.jpg

https://www.bbc.co.uk/news/uk-england-37320399

https://www.bbvaopenmind.com/en/science/leading-figures/james-lind-and-
scurvy-the-first-cl

https://www.sciencehistory.org/files/dm-scurvy-prospectorsjpg

https://commons.wikimedia.org/wiki/File:Scurvy;_male_figure._Wellcome_M
0002829.jpg

https://www.primehealthchannel.com/wp-content/uploads/2013/04/Scurvy-
Image.b197b0.webp

http://www.epi.umn.edu/cvdepi/wp-content/uploads/2011/05/Hales-Horse.jpg

https://allthatsinteresting.com/botfly-larvae

https://allthatsinteresting.com/botfly-larvae

https://commons.wikimedia.org/wiki/File:D._hominis_adult_female._png

https://allthatsinteresting.com/botfly-larvae

https://bugsinourbackyard.org/the-human-botfly-dermatobia-hominis/

https://allthatsinteresting.com/botfly-larvae

https://www.geograph.org.uk/photo/2510649

https://upload.wikimedia.org/wikipedia/commons/2/27/Giant_hogweed%2C_
Minnowburn%2C_Belfast_-_geograph.org.uk_-_3053893.jpg

https://www.zmescience.com/science/news-science/giant-hogweed-toxic-plant-
3243/

https://wellcomecollection.org/works/x5b5q784/images?id=kev3tx3m

www.rmg.co.uk/stories/blog/curatorial/removing-bladder-stone-size-tennis-ball

https://www.loc.gov/item/94505148/

https://commons.wikimedia.org/wiki/File:St_Martin_Alexis.jpg

https://commons.wikimedia.org/wiki/File:Paraponera_clavata_(14500014836).jpg

https://www.bing.com/images/search?view=detailV2&ccid=q35CPHDI&id=550B8D7EC2AE87DA6BE53FAF394437C8268AB592&thid=OIP.q35CPHDILAjhYGJv8u3YRQHaIO&mediaurl=https%3a%2f%2fhips.hearstapps.com%2fesq.h-cdn.co%2fassets%2f15%2f33%2f1439482799-justin-schmidt-1.jpg%3fresize%3d480%3a*&cdnurl=https%3a%2f%2fth.bing.com%2fth%2fid%2fR.ab7e423c70c82c08e160626ff2edd845%3frik%3dkrWKJsg3RDmvPw%26pid%3dImgRaw%26r%3d0&exph=533&expw=480&q=justin+schmidt+and+bullet+ant&simid=608021907788413059&FORM=IRPRST&ck=4B0B26EFCB3BCB2D12E25C548B2511E6&selectedIndex=0&idpp=overlayview&ajaxhist=0&ajaxserp=0

https://commons.wikimedia.org/w/index.php?search=bullet+ant+and+stinger&title=Special:MediaSearch&go=Go&type=image

https://www.troab.com/amazon-tribes-painful-rituals-to-prove-adulthood/

https://commons.wikimedia.org/w/index.php?fulltext=Search&search=File%3BScorpion+Photograph+By+Shantanu+Kuvesk&title=Special:Search&ns0=1&ns6=1&n

https://commons.wikimedia.org/wiki/File:Emperor_Scorpion_in_hand_John.jpg

https://commons.wikimedia.org/wiki/File:Parabuthus_transvaalicus_(male).jpg

https://factzoo.com/book/deathstalker-yellow-poison-stinger-desert/

https://commons.wikimedia.org/wiki/File:Avispa_marina_cropped.png

https://commons.wikimedia.org/wiki/File:Box_jellies_over_sand_at_Geldkis_DSC00331.JPG

https://wetsuitwarehouse.com.au/collections/stinger-suits/products/adrenalin-adults-hooded-lycra-stinger-suit

https://en.wikipedia.org/wiki/File:Marinesting1.jpg

https://boxjellyfish.org/australian-box-jellyfish-facts-natures-deadliest-creature/

http://imgkid.com/box-jellyfish-sting-scars.shtml

https://commons.wikimedia.org/wiki/File:Hapalochlaena_fasciata_Toba1.jpg

https://commons.wikimedia.org/wiki/File:Blue_Ringed_Octopus.png

https://commons.wikimedia.org/wiki/File:Hapalochlaena_lunulata2.JPG

https://www.animalspot.net/wp-content/uploads/2015/05/Blue-Ringed-Octopus-Size.jpg

https://commons.wikimedia.org/wiki/File:Goldenergiftfrosch1cele4.jpg

https://commons.wikimedia.org/wiki/File:Yellow-banded.poison.dart.frog.arp.jpg

https://commons.wikimedia.org/wiki/File:Ameerega_trivittata_(Madre_de_Dios,_Peru).jpg

https://commons.wikimedia.org/wiki/File:Strawberry_poison-dart_frog_(Oophaga_pumilio_or_Dendrobates_pumilio)_(9429685305).jpg

https://commons.wikimedia.org/wiki/File:Red_poison_dart_frog_(cropped).jpg

https://commons.wikimedia.org/wiki/File:Erythrolamprus_epinephalus_32731632.jpg

https://commons.wikimedia.org/wiki/File:Dendrocnide_moroides_foliage_SF20326.jg

https://www.snexplores.org/article/australian-stinging-tree-touch-pain-toxin-gympie-gymp

https://www.snexplores.org/article/australian-stinging-tree-touch-pain-toxin-gympie-gymp

https://allthatsinteresting.com/gympie-gympie

https://www.science.org/doi/10.1126/sciadv.abb8828

https://commons.wikimedia.org/wiki/File:Red-legged_Pademelon.jpg

https://www.abc.net.au/news/2021-05-25/mawson-ill-fated-far-eastern-partys-antarctic-voyage-memorial/100161834

https://collection.sl.nsw.gov.au/record/Yr86777n – viewer

https://factsforantarctica.weebly.com/map-of-mawsons-journey-in-antarctica.html

https://commons.wikimedia.org/wiki/File:Douglas_Mawson_recuperating. jpg

https://commons.wikimedia.org/wiki/File:Atropa_bella-donna_sl5.jpg

https://commons.wikimedia.org/wiki/File:20190830Atropa_belladonna1. jpg

https://commons.wikimedia.org/wiki/File:Eye_treated_with_dilating_eye_drops.jpg

https://commons.wikimedia.org/wiki/File:Nepenthes_rajah.png

https://www.bing.com/images/search?view=detailV2&ccid=yBFsOesX&id=182FEE27733FEA02842609DA22CA9E41ABA9EE56&thid=OIP.yBFsOesXfRINpidzYdUcyQHaLH&mediaurl=https%3a%2f%2fallthatsinteresting.com%2f

wordpress%2fwp-content%2fuploads%2f2018%2f04%2fpitcher-plant-lizard.jpg&cdnurl=https%3a%2f%2fth.bing.com%2fth%2fid%

https://www.ripleys.com/weird-news/shrew-poo-plant/?ref=prevpost

https://commons.wikimedia.org/wiki/File:Hyla_cinerea_pitcher_plant.jpg

https://commons.wikimedia.org/wiki/File:Dionaea_muscipula_Royal_Red_Venus_Fly_Trap.jpg

https://commons.wikimedia.org/wiki/File:Venus_Flytrap_showing_trigger_hairs.jpg

https://commons.wikimedia.org/wiki/File:Dionea_in_action.jpg

https://commons.wikimedia.org/wiki/File:Dionaea,_muscoid_fly.jpg

https://commons.wikimedia.org/wiki/File:Bladderwort_(Utricularia)_in_bloom,_Cass_County,_Texas,_USA_(April_2017).jpg

https://commons.wikimedia.org/wiki/File:Common_Bladderwort_(3629471561).jpg

https://commons.wikimedia.org/wiki/File:Utricularia_vulgaris_001.JPG

https://commons.wikimedia.org/wiki/File:Utricularia_aurea_8_Darwiniana.jpg

https://carnivorousplantresource.com/the-plants/bladderwort/

https://commons.wikimedia.org/wiki/File:Drosera_rotundifolia_Rosiczka_okr%C4%85g%C5%82olistna_2022-06-21_06.jpg

https://commons.wikimedia.org/wiki/File:Drosera_anglica_ne2.jpg

https://commons.wikimedia.org/wiki/File:Round-leaved_Sundew_(Drosera_rotundifolia)_with_ant_prey_(9410495122).jpg

https://commons.wikimedia.org/wiki/File:Drosera_capensis_bend.JPG

https://commons.wikimedia.org/wiki/File:Pulex_irritans_ZSM.jpg

https://commons.wikimedia.org/wiki/File:Female_human_head_louse.jpg

https://commons.wikimedia.org/wiki/File:Ixodes_scapularis.jpg-/media/File:Ixodes_scapularis.jpg

https://www.realclearscience.com/lists/gutwrenching_pictures_of_parasitic_diseases/pinworm.html - !

https://commons.wikimedia.org/wiki/File:Sarcoptes_scabei_2.jpg
NOT Available

https://commons.wikimedia.org/wiki/File:Fleabite.JPG

https://www.homenaturalcures.com/head-lice-symptoms-causes/

https://commons.wikimedia.org/w/index.php?search=lyme+disease+bite&title=Special:MediaSearch&go=Go&type=image

https://pharmamum.com/worms-in-children/

https://commons.wikimedia.org/wiki/Category:Scabies -
/media/File:ScabiesD08.JPG

https://commons.wikimedia.org/wiki/File:2015.11.26.192558_Cimex_lectulari
us_bites.jpg - /media/File:2015.11.26.192558_Cimex_lectularius_bites.jpg

https://www.dailymail.co.uk/news/article-9397203/Thai-man-excretes-59-
FOOT-tapeworm-doctors-visit-extreme-flatulence.html

https://eyewiki.org/File:Loa_loa_external.jpg

https://commons.wikimedia.org/wiki/File:Elephantiasis_of_the_leg_(ATED_74
-6426-2),_National_Museum_of_Health_and_Medicine_(283701106).jpg

https://commons.wikimedia.org/wiki/File:Aedes_aegypti_CDC8936.tif

http://www.forestryimages.org/browse/detail.cfm?imgnum=5390481

https://plantdoctor.co.nz/assets/uploads/2016/05/nematode-fungal-hyphae.jpg

https://commons.wikimedia.org/wiki/File:Cordyceps_matou_lagarta.jpg

https://commons.wikimedia.org/wiki/File:Grasshopper_cordyceps_(198852840
90)

https://commons.wikimedia.org/wiki/File:Ant_Killed_by_Fungus_-
_Cockscomb_Wildlife_Sanctuary,_Belize.jpg

https://commons.wikimedia.org/w/index.php?search=mountain+chicken+frog
&title=Special:MediaSearch&go=Go

https://commons.wikimedia.org/wiki/File:Chytridiomycosis.jpg

https://www.livescience.com/19628-fungal-diseases-emerging-threat.html.

www.ingramcontent.com/pod-product-compliance
Lightning Source LLC
Chambersburg PA
CBHW041647200526
45172CB00022BA/1287